GARETH STEVENS
VITAL SCIENCE
Physical Science

THE NATURE OF MATTER

by Anna Claybourne
Science curriculum consultant: Suzy Gazlay, M.A.,
science curriculum resource teacher

GARETH STEVENS
GS
PUBLISHING
A Member of the WRC Media Family of Companies

Please visit our Web site at: www.garethstevens.com
For a free color catalog describing Gareth Stevens Publishing's list of high-quality books and multimedia programs, call 1-800-542-2595 (USA) or 1-800-387-3178 (Canada).
Gareth Stevens Publishing's fax: (414) 332-3567.

Library of Congress Cataloging-in-Publication Data

Claybourne, Anna.
The nature of matter / by Anna Claybourne.
p. cm. — (Gareth stevens vital science. Physical science)
Includes bibliographical references and index.
ISBN-13: 978-0-8368-8088-5 (lib. bdg.)
ISBN-13: 978-0-8368-8097-7 (softcover)
1. Matter. I. Title.
QC171.2.C55 2006
530—dc22 2006033732

This edition first published in 2007 by
Gareth Stevens Publishing
A Member of the WRC Media Family of Companies
330 West Olive Street, Suite 100
Milwaukee, WI 53212 USA

Produced by Discovery Books
Editors: Rebecca Hunter, Amy Bauman
Designer: Melissa Valuch
Photo researcher: Rachel Tisdale

Gareth Stevens editorial direction: Mark Sachner
Gareth Stevens editors: Carol Ryback and Barbara Kiely Miller
Gareth Stevens art direction: Tammy West
Gareth Stevens graphic design: Scott Krall
Gareth Stevens production: Jessica Yanke and Robert Kraus

Illustration by Stefan Chabluk
Photo credits: Adidas: p. 35; CFW Images: pp. 4 (Edward Parker/EASI-Images), 5 (Neal Cavalier-Smith/EASI-Images), 7 (Simon Scoones/EASI-Images), 18 (Dawne Fahey/EASI-Images), 21 (Clive Sanders/EASI-Images), 36 (Edward Parker); Getty Images: p. 10 (Lester Lefkowitz); Istockphoto: pp. 15 (Klaas Lingbeek-van Kranen), 22 (Paul-Andre Belle-Isle), 24 (Lucian Coman), 27 (Pierangelo Rendina), 30 & title page (Tomislav Forgo), 32 (Udo Weber); Library of Congress: p. 9; MCAS: p. 39 (Lance Cpl. Robert W. Beaver); NASA: pp. 38 (JPL), 42, 43. Cover:

Printed in Canada

1 2 3 4 5 6 7 8 9 10 10 09 08 07 06

TABLE OF CONTENTS

Words that appear in the glossary are printed in **boldface** type the first time they appear in the text.

Cover: Water exists in all three states of matter, two of which, solid and liquid, can be seen here.

Title page: Water can be made into ice when the temperature is at or below freezing.

Introduction

What is matter? It simply means any material from which anything physical is made. Matter has substance, and it takes up space. Everything in the universe—from planets and stars to rocks, cars, tables, chairs, plants, people, air, water, and food—is made of matter.

Physical things

Things that are made of matter are physical things. That means they are real, three-dimensional (3-D) things you can feel, such as metal, wood, water, and air. (Not everything is made of matter. For example, heat and light are forms of energy.) The science of matter and **energy** and how they interact is called **physics**, and a scientist who studies matter is a physicist.

Everything in the universe is made of matter, from giant planets to tiny specks of dust. Earth and all the things on it, including rocks, humans, animals, and plants are made of matter too.

Elements

There are more than one hundred different basic types of matter, called **elements**. Gold, iron, and oxygen are elements. Different elements can combine together to make thousands more types of matter.

Atoms

Elements are made up of tiny parts called **atoms**. An atom is the smallest single part of an element, containing tiny **particles** around a central **nucleus**. Atoms can join together with other atoms of the same type or atoms of different types to make **molecules** of different substances. The science of atoms, their **properties**, and how they join together and make substances is called **chemistry**.

Material world

Different types of matter are often called materials. There are many thousands of different types of matter. Look around you now, and you'll probably see dozens of different materials: rubber, wood, plastics, paper, metals, fabrics, and stone.

We use different materials for different jobs, depending on how they behave. For example, steel is strong and flexible, so it's used in buildings and bridges. Plastic does not conduct electricity, so it's good for screwdriver handles.

Matter matters

This book is about how matter works. It begins with the atom, explaining what atoms are, how they bond to make materials, and how these materials behave. Along the way, you will find answers to questions such as these: Just how tiny is an atom? Are atoms made of even smaller parts? Why is gold so heavy? What makes a car get rusty? How can eating food give us energy?

Knowing some of this information can help us understand ourselves, our world, and our universe.

MADE OF WATER?

"Water is the principle, or the element, of things. All things are water."

Ancient Greek philosopher Thales of Miletus (c.625–c.546 B.C.) believed that matter came to be from water. Although his theory was wrong, it was revolutionary compared to the beliefs at the time.

Everything in the world, including all plants, animals, and people is made of matter.

Did You Know ?

When some people talk about "chemicals," they may mean something added to their food or sprayed onto crops. But in fact, all materials are chemicals. A chemical is simply a substance made up of atoms.

Atoms

Matter is made up of tiny particles called atoms. A single atom is much too small to be seen with the naked eye. But when you look at matter in the form of an object—for example a wooden desk, a coin, or your own hand—you're looking at matter made up of trillions and trillions of atoms.

How small *is* an atom?

Atoms are unbelievably small. They come in different sizes, but the average size of an atom is about one ten-millionth of a millimeter across (0.0000001 mm). In inches, that's one 250-millionth of an inch (0.000000004 inches).

In other words, one page of this book is about one million (1,000,000) atoms thick.

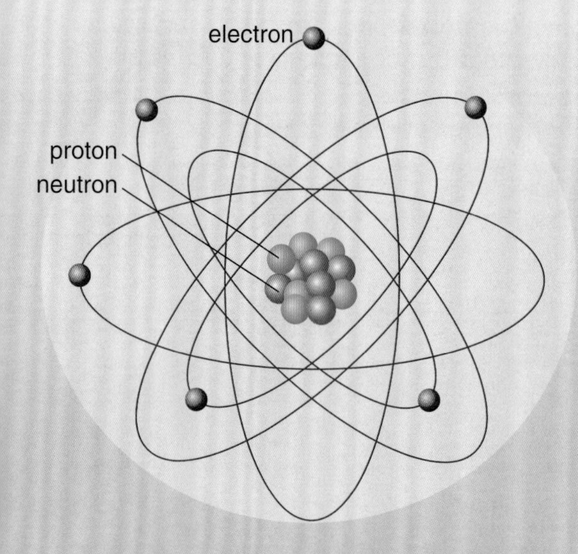

This diagram shows the structure and parts of an atom. This is not what an atom actually looks like. In real life, the electrons move so fast it is very hard to see them, and the nucleus is much smaller compared to the cloud of electrons around it.

Atomic structure

Over the last two hundred years, scientists have performed many experiments to figure out what is inside an atom.

Atoms are made up of smaller particles, which are often referred to as **subatomic particles**. Each atom has a central part called a nucleus. The nucleus contains two types of tiny subatomic particles: positively charged **protons** and neutrally charged **neutrons**. Some even smaller particles found in the atom—the negatively charged

EVERYTHING IS ATOMS

"Nothing exists except atoms and empty space; everything else is opinion."

Greek philosopher Democritus (c.460–370 B.C.), the first person to suggest that everything is made of atoms and the first person to use the word "atom", which means "indivisible" in Greek.

electrons—rapidly orbit the nucleus in specific paths.

Electron shells

Different types of atoms have different numbers of electrons. They move around the nucleus in layered paths, known as electron shells. The innermost shell can hold up to two electrons, and the next two shells can contain as many as eight electrons each. Outer shells can hold even more electrons. Each atom has as many shells as it needs to hold all its electrons. Some atoms have as many as seven shells.

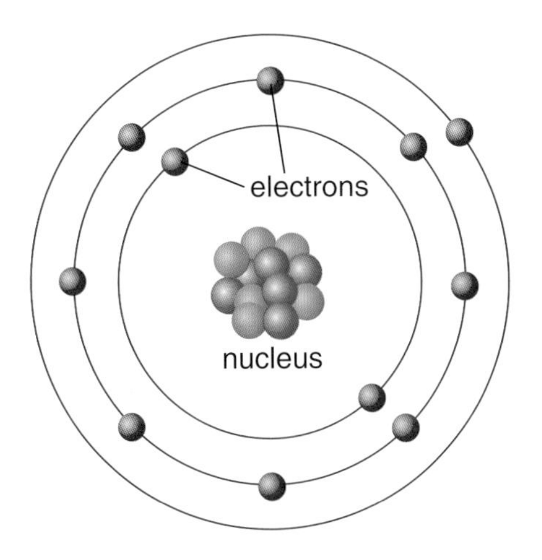

A sodium atom has eleven electrons: Two in the first shell, eight in the second shell, and one in the third shell.

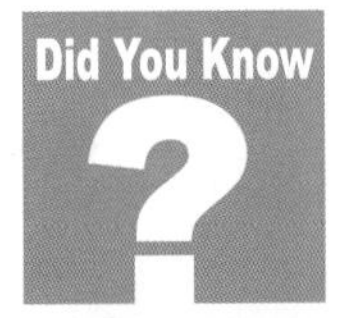

Gold leaf is one of the thinnest materials you can find in everyday life. It is one of several metals that can be hammered into thin foil-like sheets, which are most often used for decorative purposes. A single sheet is about three hundred atoms thick.

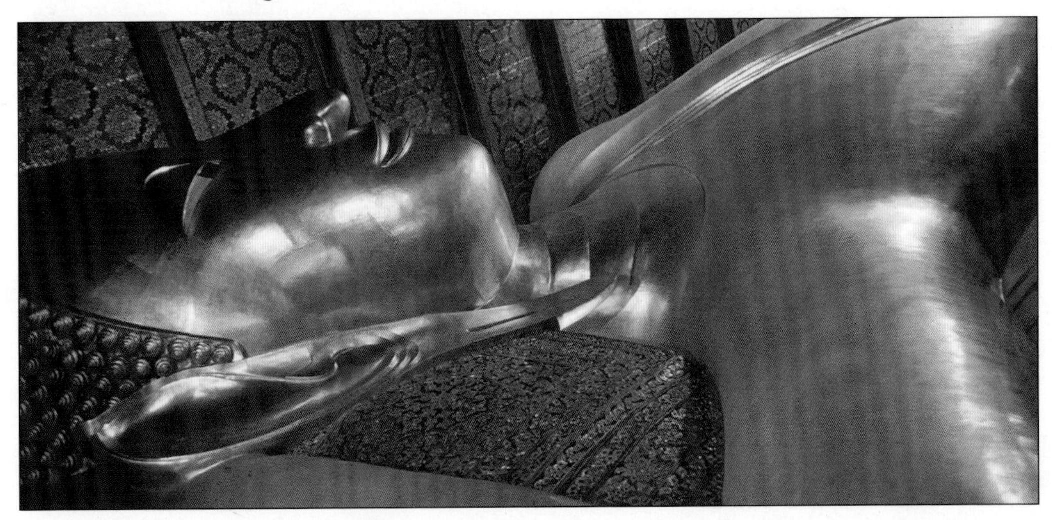

The famous *Reclining Buddha* statue at Wat Pho temple in Thailand is covered with a layer of gold leaf.

Properties of atoms

The properties of an atom are the characteristics of it that we can measure and use to describe it. First, each atom has its own **atomic number** and **mass number**. These are determined by the number of protons and neutrons in its nucleus.

Atomic number

The atomic number of an atom is equal to the number of protons in its nucleus. For example, an oxygen atom has eight protons in its nucleus, so its atomic number is 8.

Mass number

An atom's mass number is equal to the total number of protons and neutrons in its nucleus. For example, an oxygen atom has eight protons and eight neutrons. So its mass number is 8 + 8, or 16.

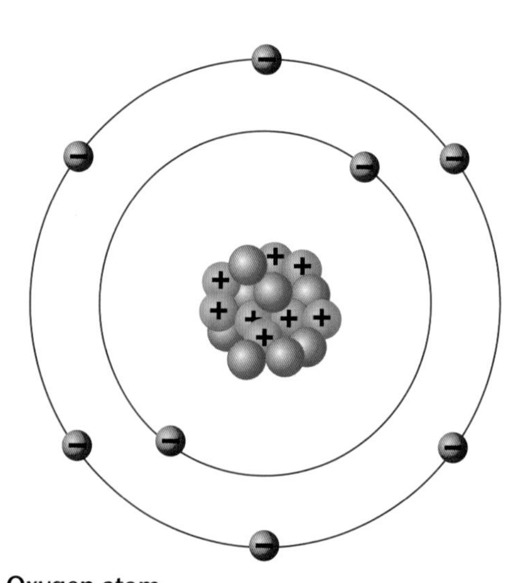

Oxygen atom nucleus

8 protons
Atomic number = 8

8 protons + 8 neutrons
Mass number = 16

Electrical charge

Atoms are held together by **electric charge**. Protons have a positive electric charge, written as a plus sign (+). Electrons have a negative electric charge, written as a

WRITING IT DOWN

Each atom has a special symbol that stands for its name. For example, the symbol for an oxygen atom is O, and the symbol for a sodium atom is Na. We can write down information about an atom in a simple pattern that looks like this:

$$^{16}_{8}\mathrm{O}$$

The mass number is written to the top left of the symbol. The atomic number is written to the bottom left of the symbol. The middle letter (or letters) is the symbol for the element, in this case oxygen.

Electrical charge inside an atom

Protons are positive (+)

Electrons are negative (-)

Neutrons are neutral

minus sign (-). Positive and negative charges are attracted to each other and pull toward each other. Neutrons are neutral—they have no electrical charge.

Usually, an atom has the same number of protons and electrons. For example, the oxygen atom pictured opposite has eight protons and eight electrons. They balance each other out, so overall the atom has no electric charge.

Elements

An element is a substance that is made up of a single type of atom. Because of this, elements are said to be "pure" substances. This also means they cannot be broken down into other substances. Thus, the element oxygen contains oxygen atoms and nothing else; the element gold contains only gold atoms, and so forth.

So far, we know of more than one hundred basic types of atoms, or elements.

MASS AND CHARGE

Protons, neutrons, and electrons have **mass.** This means they contain a certain amount of matter. The mass number of an individual atom is a measurement of its mass. Protons and neutrons each have one **atomic mass unit**, or amu. Electrons have very little mass and are not counted.

Particle	Mass	Charge
Proton	1	+1
Neutron	1	0
Electron	0.0005	-1

JOHN DALTON

John Dalton (1766–1844) was a schoolteacher from Cumbria, England, who studied matter and chemicals in his spare time. He realized that different substances were made up of different types of atoms and that each atom has its own properties. Dalton could not see atoms, but he came up with a **theory**—an explanation based on careful study—to explain why different substances behaved as they did. Because he saw that substances combined and behaved in different ways, he reasoned that they must be made up of different atoms with different properties. Modern microscope pictures of atoms prove he was right. Dalton's ideas form the basis of modern matter science, and because of his ideas, he is often called "the father of chemistry." The unit of measurement known as the atomic mass unit, or amu, is also known as a Dalton, or Da, and is named for John Dalton.

Eighty-eight of these occur naturally on Earth, although there is some disagreement among scientists about that number. The others are **synthetic**, which means that they are made artificially. Technetium (43) and promethium (61) are created in a cyclotron, also known as a particle accelerator or "atom smasher." All the elements that come after uranium (92) are made in nuclear reactors. The known elements can be represented in a chart known as the **Periodic Table of Elements** (see page 12). In the Periodic Table the elements are numbered in order beginning with 1 (hydrogen). Most representations of the Periodic Table have room for 118 elements, even though all 118 have not yet been discovered. Researchers continue to try to create additional elements, so the number of known elements is likely to keep changing.

Where do the elements get their names? Some are named after scientists, and some are named for places where they were discovered. Uranium, neptunium, and plutonium were named in order after the planets. (Pluto was classified as a planet at the time!) Antimony, which was used as eye makeup in ancient Rome, gets its name from a word meaning "cosmetic." Cobalt comes from a German word for "gremlin," because miners thought cobalt was worthless and corrupted the ore they were mining. If you look them up, you will find that nearly every name has an interesting history.

KEEP LOOKING...

"Nothing, from mushrooms to a scientific law, can be discovered without looking and trying. So I began to look about and write down the elements with their atomic weights and typical properties..."

Russian scientist Dmitry Mendeleyev (1834–1907), who invented the Periodic Table of the Elements.

Gold is an element. These gold bars are made up of only one type of atom—gold atoms.

ISOTOPES

Isotopes are variations of the same atom. They have the same number of protons as the other atoms of the same name, but each isotope has a different number of neutrons. Carbon, for example, has three isotopes known as carbon-12, carbon-13, and carbon-14. Each of these atoms has six protons and an atomic number of six. But because they have different numbers of neutrons, each has a different mass number.

$^{12}_{6}C$	Carbon-12 6 protons 6 neutrons Mass number: 12	$^{13}_{6}C$	Carbon-13 6 protons 7 neutrons Mass number: 13	$^{14}_{6}C$	Carbon-14 6 protons 8 neutrons Mass number: 14

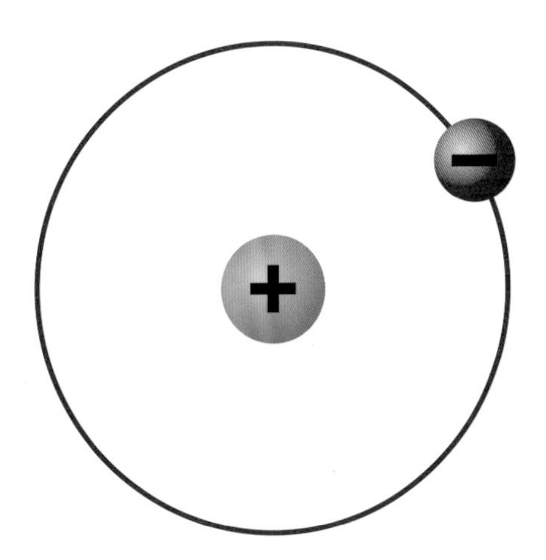

Hydrogen atom
Atomic number: 1
1 proton
0 neutrons
1 electron

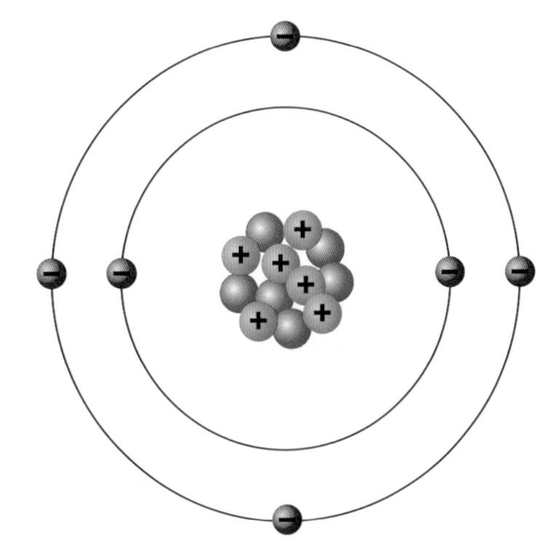

Carbon atom
Atomic number: 6
6 protons
6 neutrons
6 electrons

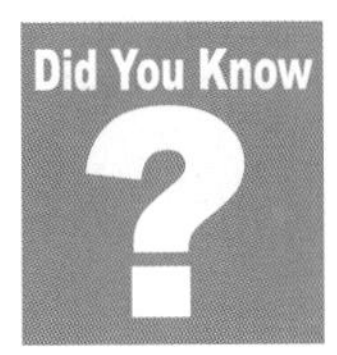

Long ago, people used to think that matter was made up of just four elements: earth, air, fire, and water. In modern science, we know that none of these is an element. All of them are made of substances containing multiple types of atoms.

The Periodic Table of the Elements

The Periodic Table of the Elements shows all of the elements arranged according to their properties. Its name is often shortened to simply the Periodic Table. The table, which was compiled by Russian chemist Dmitry Mendeleyev (1834–1907) in 1869, presents the elements in a few different ways. First, the elements are arranged by atomic number, starting with hydrogen, which has an atomic number of 1. Then, the table lists the elements by common properties. Grouping them this way presents the elements in rows, called **periods**, and columns, called **groups**. Each of these "features" of the chart tells us something more about the elements.

Periods

The rows, or periods (from which the Periodic Table gets its name), correspond to the number of electron shells an atom

The Periodic Table of the Elements

							H 1										He 2
Li 3	Be 4											B 5	C 6	N 7	O 8	F 9	Ne 10
Na 11	Mg 12											Al 13	Si 14	P 15	S 16	Cl 17	Ar 18
K 19	Ca 20	Sc 21	Ti 22	V 23	Cr 24	Mn 25	Fe 26	Co 27	Ni 28	Cu 29	Zn 30	Ga 31	Ge 32	As 33	Se 34	Br 35	Kr 36
Rb 37	Sr 38	Y 39	Zr 40	Nb 41	Mo 42	Tc 43	Ru 44	Rh 45	Pd 46	Ag 47	Cd 48	In 49	Sn 50	Sb 51	Te 52	I 53	Xe 54
Cs 55	Ba 56	57-71	Hf 72	Ta 73	W 74	Re 75	Os 76	Ir 77	Pt 78	Au 79	Hg 80	Tl 81	Pb 82	Bi 83	Po 84	At 85	Rn 86
Fr 87	Ra 88	89-103	Rf 104	Db 105	Sg 106	Bh 107	Hs 108	Mt 109	Ds 110	Rg 111	Unb 112	Uut 113	Uuq 114	Uup 115	Uuh 116	Uus 117	Uuo 118

La 57	Ce 58	Pr 59	Nd 60	Pm 61	Sm 62	Eu 63	Gd 64	Tb 65	Dy 66	Ho 67	Er 68	Tm 69	Yb 70	Lu 71
Ac 89	Th 90	Pa 91	U 92	Np 93	Pu 94	Am 95	Cm 96	Bk 97	Cf 98	Es 99	Fm 100	Md 101	No 102	Lr 103

Metals Metalloids Non-metals Undiscovered

The Periodic Table often appears color-coded. The colors tell us something else about an element—in this case, the type of element it is.

has. The first period contains atoms with only one electron shell. As the first electron shell can hold up to two electrons, there are only two atoms in this period. The next period contains atoms with two electron shells, and so on.

Groups

Atoms in the same column, or group, have the same number of electrons in their outer shell. For example, Group I contains atoms with one electron in their outer shell. The atoms in each group also have similar chemical properties.

As you can see, most elements are metals. Metals are elements that are good at conducting electricity, and most of them react easily with other atoms to form new substances.

Two rows of elements are shown separately from the rest of the tables. The first consists of all the elements from 57 (lanthanum) to 71 (lutetium). These are called the lanthanides, also known as "rare earth" or "inner transition" elements, and they are part of Period 6. The second row includes the elements from 89 (actinium) to 103 (lawrencium) and is part of Period 7. These elements are the actinides, and they are all radioactive. Some of them are not even found in nature. Instead, they have been made in labs. The ability to make elements makes it possible that elements can continue to be added to the Periodic Table, beyond any that have been discovered in nature.

READING THE TABLE

The Periodic Table provides a lot of information about the elements. Each element is represented by a box that contains its name, symbol, atomic number, and RAM, or **relative atomic mass.** The RAM represents the average mass of an atom of that element. It is often the same as the mass number. (The table on page 12 shows each element's symbol and atomic number.)

8	**Atomic number**
0	**Symbol**
Oxygen	**Name of element**
16.0	**Relative atomic mass (RAM)**

Hydrogen is the most common element in the universe. The most common element in Earth's crust is oxygen.

Molecules

Atoms don't just exist by themselves. They join together, or bond, with other atoms to make larger units of matter, called molecules. Most types of matter are made up of molecules rather than one type of atom. For example, salt molecules are made of sodium and chlorine atoms bonded together. Sugar is a combination of bonded carbon, hydrogen, and oxygen atoms.

Molecule types

A molecule is the smallest particle of a substance that has all the characteristics and properties of that substance. It can be an **elemental molecule**—one or more of the same types of atoms joined together. Or it can be a **compound** molecule—a bonded combination of different types of atoms.

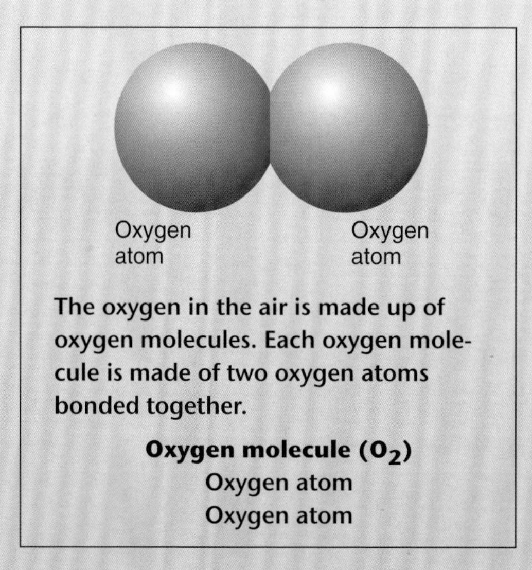

The oxygen in the air is made up of oxygen molecules. Each oxygen molecule is made of two oxygen atoms bonded together.

Oxygen molecule (O_2)
Oxygen atom
Oxygen atom

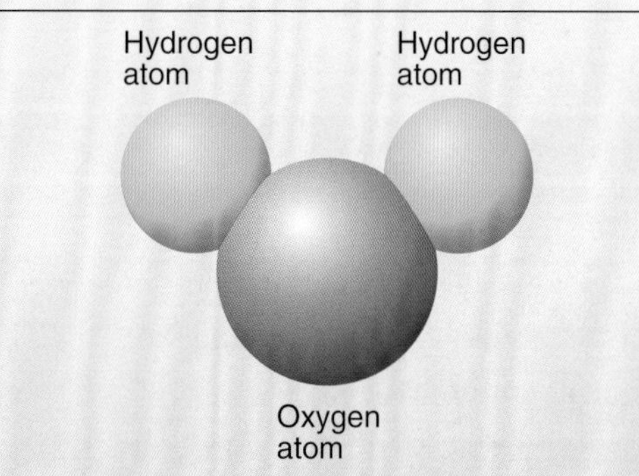

A water molecule is made up of two hydrogen atoms bonded to one oxygen atom.

Water molecule (H_2O)
Oxygen atom
Hydrogen atom
Hydrogen atom

The H shows that water contains hydrogen. The O shows that water contains oxygen. The 2 after the H shows there are two atoms of hydrogen for every atom of oxygen.

Each type of molecule has unique characteristics. It is totally different from any of the atoms from which it is made. For example, sodium—the element—is a soft, silvery metal that reacts violently if it comes in contact with water. When that happens, it releases heat and hydrogen **gas**, forms a corrosive solution called sodium hydroxide, and may even explode. Chlorine is a dense, greenish gas that smells terrible and is poisonous to

MOLECULAR MODELS

Scientists and researchers use diagrams and models to show how atoms join together to make up molecules. Two of the most common models are the space-filling models and the ball-and-stick models. Space-filling models show ball-like atoms clumped together to make molecules. These models come close to showing what the structure of molecules is like. Ball-and-stick models show ball-like atoms linked by sticks to show the bonds between them. This type of model can show some complicated molecules more clearly.

The ball-and-stick model (on the right) and the space-filling model (on the left) are two different ways to show a methane (CH_4) molecule.

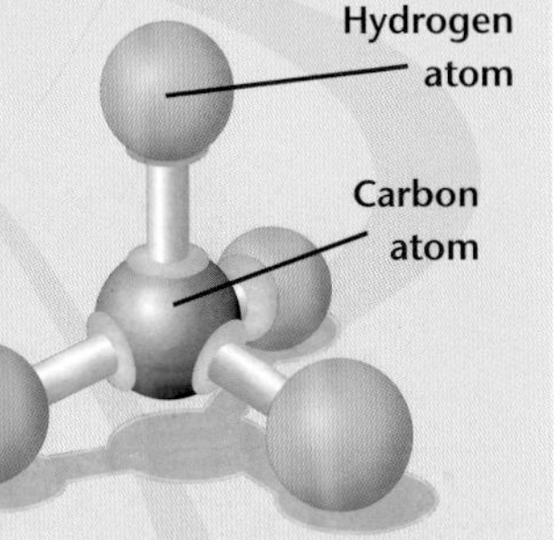

anyone who breathes it. Clearly, these two elements are not substances that anyone would want to have around! Yet, when atoms of sodium and chlorine bond together to form a molecule, the result is a substance that you probably eat every day: sodium chloride (NaCl), commonly known as table salt.

Water is made up of millions of microscopic water molecules.

CHEMICAL FORMULAS

A molecule can be written down as a chemical formula, which is a kind of code or symbol made up of letters and numbers. The letters show what types of atoms are in the molecule. The numbers show how many atoms of each kind are included. For example, the chemical formula for water is H_2O and the chemical formula for sucrose (white sugar) is: C_{12} H_{22} O_{11}

Why bonding happens

Bonding is the way that atoms join with other atoms to make molecules. So, why do atoms bond? It has to do with the layers of electrons surrounding an atom's nucleus (see page 7).

Each electron shell has a maximum number of electrons that it can contain. This number differs from shell to shell. Although, as noted, the innermost electron shell can hold only two, outer shells can hold up to thirty-two electrons. If an atom's outer electron shell is filled with electrons, the atom is said to be stable. That means it is strong and can exist easily by itself. Interestingly, outer shells act "full" with as few as eight electrons. If the outer shell is not full, however, the atom is said to be unstable. When an atom is unstable, it may join with other atoms, which will change its number of electrons and help make the atom more stable.

Ionic bonding

One type of bonding is ionic bonding. This type of bonding happens when an atom loses or gains electrons in bonding with another atom. The electrons are completely transferred from one atom to the other.

Ionic bonding

Sodium atom
Sodium (Na) is a metal that has one electron in its outer shell.

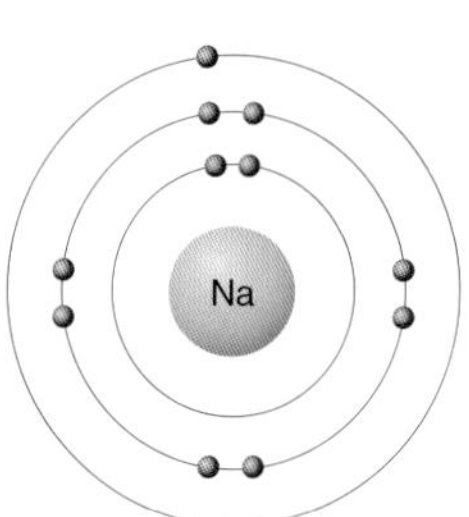

Sodium ion
The sodium atom loses an electron to stabilize itself. Because electrons have a negative charge, the sodium atom now has a positive charge.

Chlorine atom
Chlorine (Cl) is a nonmetal element that has seven electrons in its outer shell. It needs another to become stable.

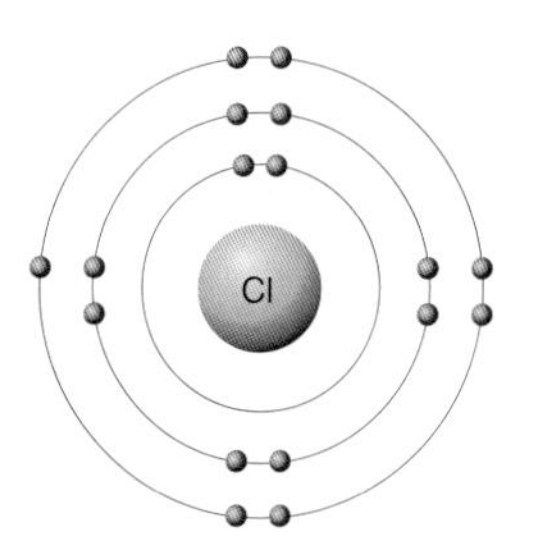

Chlorine ion
The chlorine atom collects a spare electron from the sodium atom to become stable. With an extra electron, it now has a negative charge.

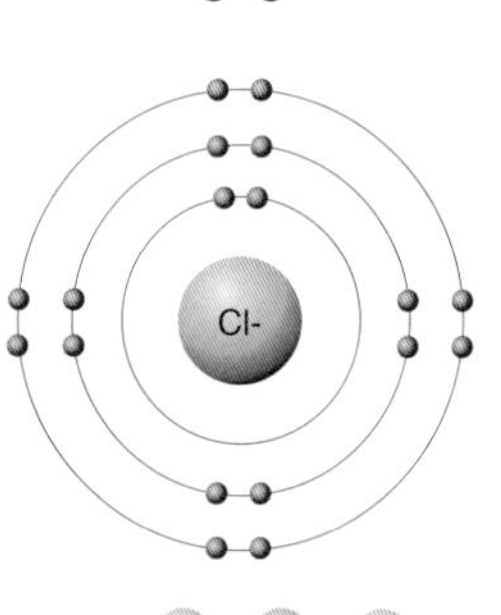

Sodium chloride molecule
A sodium atom has a positive charge and a chlorine atom has a negative charge. Because oppositely charged atoms attract one another, these two atoms are attracted to each other. Sodium and chlorine ions bond together in a grid structure to make a sodium chloride molecule (NaCl).

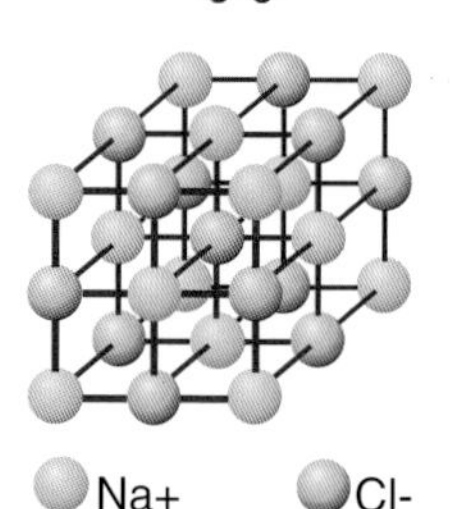

Have you ever heard the phrase, "opposites attract"? That's exactly what happens when ionic bonding takes place.

Keep in mind that electrons have a negative charge. Atoms with an electrical charge due to missing or extra electrons are called **ions**. An ion missing an electron has a positive charge, and an ion with an extra electron has a negative charge. The result is a powerful attraction between those two oppositely charged ions. They come together forming an ionic bond.

Sodium chloride, or common salt, is made by ionic bonding. The diagram on page 16 shows how it works.

Covalent bonding

Ionic bonding works well for some atoms, but others would have to gain or lose too many electrons to form ions. So, instead of transferring electrons, these atoms join together to share them. This type of bonding is called covalent bonding. Elements that are close together on the Periodic Table are more likely to form covalent bonds.

Every pair of electrons formed between atoms is a single covalent bond. Some atoms can share more than one pair of electrons, so they have multiple covalent bonds. The more electrons that are shared, the stronger the bonds. A hydrogen molecule, consisting of two hydrogen atoms, has a single covalent bond. It can be broken apart more easily than an oxygen molecule, consisting of two oxygen atoms, which has two covalent bonds. Even stronger is a nitrogen molecule, consisting of two nitrogen atoms, but with three pairs of shared electrons forming three covalent bonds.

Water molecules are an example of covalent bonding (see below).

Hydrogen atoms have one electron in their outer shell. To be stable, they need two electrons.

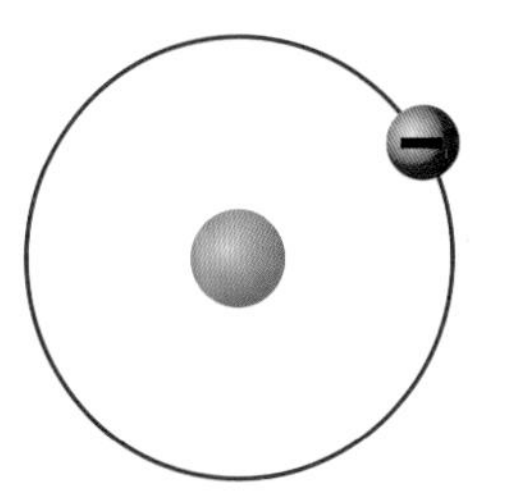

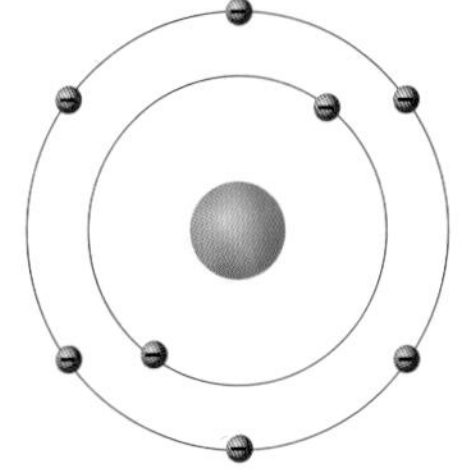

An oxygen atom has six electrons in its outer shell. To be stable, it needs eight.

A covalently-bonded water molecule

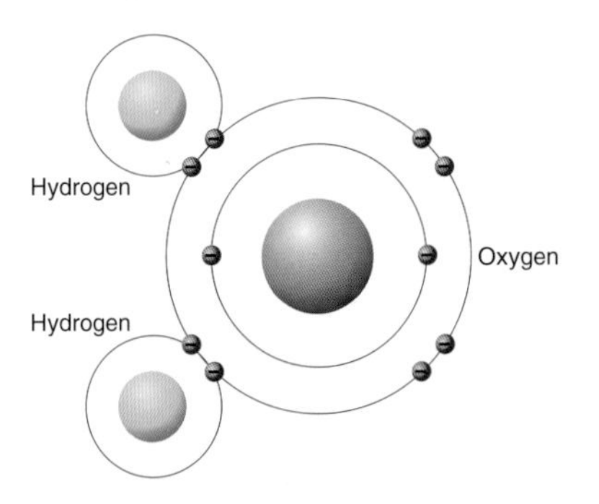

In a water molecule, one oxygen atom and two hydrogen atoms share the electrons in their outer shells. They each end up with enough electrons to become stable.

Chemical reactions

When atoms and molecules meet, they may exchange or share electrons, coming together by a chemical bond to make a new substance. Sometimes when two substances meet, however, the result is a chemical change. In this case, the two substances rearrange or swap their atoms and become different substances. In either situation, the result is called a **chemical reaction**.

Rusting is another example of chemical reaction. This old truck has rusted all over. This reaction is actually a chemical change, called an oxidation reaction. It happens when metals react with water and oxygen in the air.

Writing it down

When oxygen and hydrogen bond to make water, for example, a simple chemical reaction is happening. We record such chemical reactions using the chemical formulas for the different atoms and molecules. The water reaction (on page 19) is written like this: $2H_2 + O_2 \rightarrow 2H_2O$

In other words...
two hydrogen molecules ($2 \times H_2$)
plus
one oxygen molecule (O_2)
become two water molecules ($2 \times H_2O$)

Reactions everywhere

Chemical reactions don't just happen in chemistry labs. They are going on all the time, all around us, between molecules in the air, in food, inside plants and animals, and in all kinds of everyday substances. For example, chemical reactions known as chemical changes are happening when iron gets rusty, when a cake is baking, when plants grow, or when fruit in your fruit bowl goes bad.

ANTOINE LAVOISIER

Antoine Lavoisier (1743–1794) was a wealthy French nobleman who became a scientist. Lavoisier discovered that water was made of hydrogen and oxygen. He also studied the chemical reactions involved in burning, rusting, and breathing. He was the first scientist to state the law of conservation of matter (see page 19). Lavoisier's work was cut short at the age of fifty-one when he was beheaded during the French Revolution.

CHEMICALS IN YOUR BODY?

Chemical reactions are going on inside your body all the time. They break food molecules into atoms and turn them into the chemicals your body needs.

How hydrogen and oxygen react to make water

1\.

Hydrogen molecules

Oxygen molecule

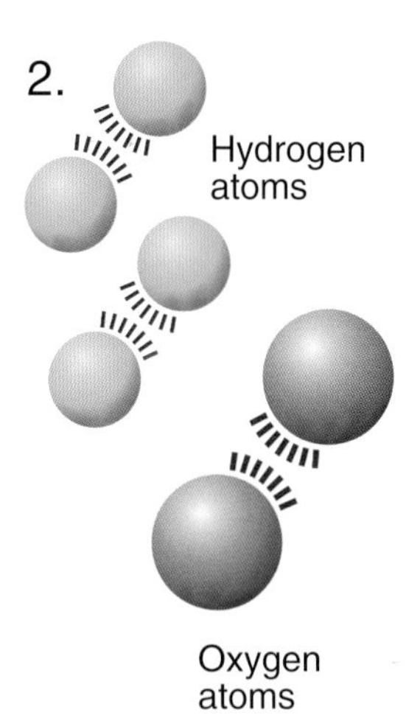

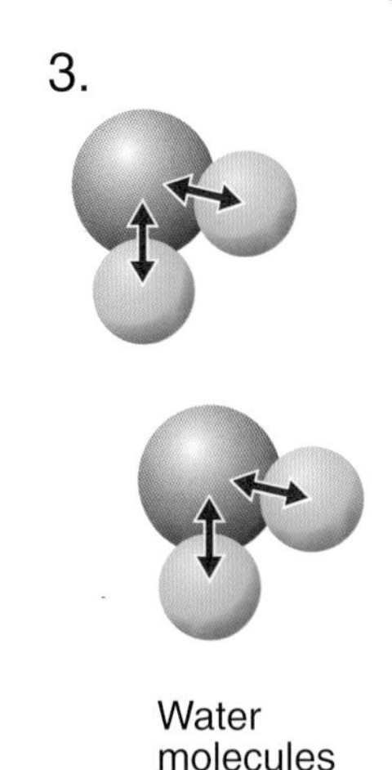

1\. Hydrogen and oxygen molecules are made of two atoms bonded together.

2\. When they react together, the bonds between the pairs of atoms break apart.

3\. The atoms bond in different patterns to form water molecules.

THE LAW OF CONSERVATION OF MATTER

When substances experience a reaction and make other substances, the total amount of matter stays the same. Before and after the reaction, the same atoms are there—they are just rearranged into different patterns. This rule is also called the "law of conservation of matter." In other words, a chemical reaction does not destroy matter and it does not create new matter. It just changes matter from one form to another.

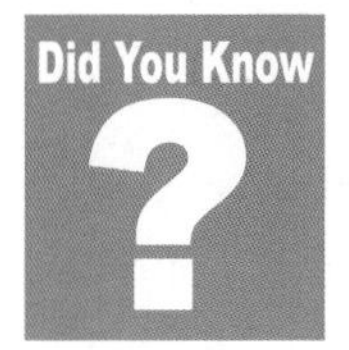

Some substances rarely react with other atoms, molecules, or substances. Such substances are stable and do not need to combine with other molecules. Gold, for example, is unreactive. Its atoms almost never bond with other atoms.

Solids, Liquids, and Gases

Every type of matter has its own set of traits, called its properties. We can use properties to distinguish one type of matter from another. One property of matter is its form, or state. There are three commonly recognized states of matter: solids, liquids, and gases. If the conditions are right, most substances can exist in each of these states, but under normal conditions, most substances will remain in a particular state.

We think of iron as a solid. If it is heated to 2,795°F (1,535°C), it will **melt** and become a liquid. At 4,982°F (2,750°C), it will begin to boil and become a vapor (gas). So, while it is possible for iron to exist in all three states, it is not commonly seen as a liquid or gas.

Water is unique in that it is the only substance that naturally occurs in all three states: solid (ice), liquid (water), and gas (water vapor). We tend to think of it as a liquid, because it is liquid at normal room temperature. We think of oxygen as a gas, because that is how it normally occurs on Earth. But it, too, can become a liquid, at a temperature of -297.2°F (-182.9°C) and a solid at -360.9°F (-218.3°C).

MOVING MOLECULES

"While examining the form of these particles immersed in water, I observed many of them very evidently in motion."

Botanist Robert Brown describes seeing for the first time what became known as Brownian motion.

Three states

As you know, solids, liquids, and gases look and feel very different from each other. This is because of the way their molecules behave, how firmly they are fixed together, and how much they move. A solid object has a definite shape. That is because in a solid, the molecules are firmly attached to each other. They **vibrate** (move back and forth a small amount) but do not leave their positions. Another characteristic of a solid is that other objects cannot pass through it.

Matter in a liquid state, however, changes shape depending upon the shape of its container. The liquid's molecules are more loosely attached to each other and can move around. Because of this, a liquid such as water or oil can run, splash, spill,

On Earth, water is the only substance that is commonly found in all three states of matter.

flow, or be poured. Also, solid objects can pass through a liquid.

In a gas, the molecules are not fixed together. They fly around in all directions at high speed. This means that a gas, like a liquid, has no fixed shape. It spreads out to fill the space it is in. Most gases are invisible, as the molecules are so widely spaced out.

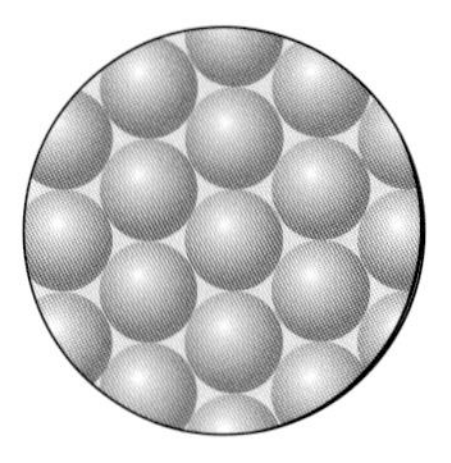

Solid
The molecules in a solid hold firmly together, so it stays the same shape.

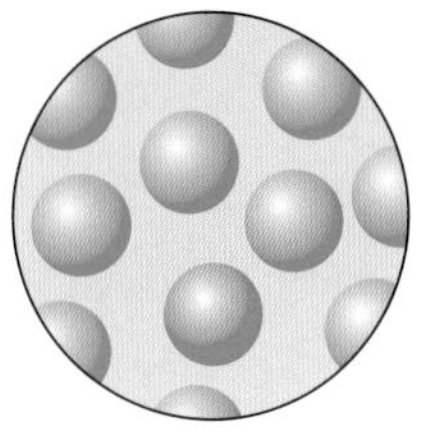

Liquid
The loose attraction between the molecules in a liquid hold it together, but it can be stirred, spilled—or even poured into a different container.

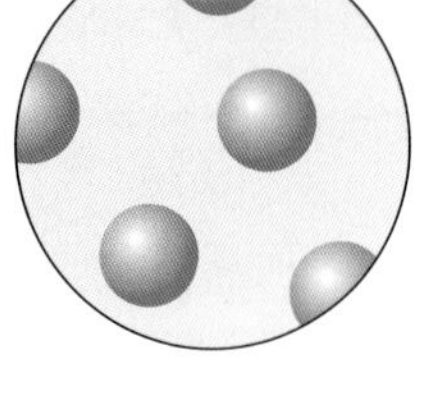

Gas
Gases are often invisible. Their molecules spread out to fill whatever container or area they are in.

ROBERT BROWN

Scottish scientist Robert Brown (1773-1858) made an important discovery about how molecules move in liquids. Oddly, Brown was not a chemist. He was a botanist, which is a scientist who studies plants. In the early 1800s, he went to Australia, where he collected thousands of plant species previously unknown to science. In 1827, he was looking at grains of pollen in water under a microscope. He saw the tiny particles jumping and bumping around in the water. It was later discovered that this movement, now called Brownian motion (or Brownian movement), is caused by the molecules moving in liquids and bumping into larger objects such as the pollen grains.

Changes of state

Many substances can change from one state of matter to another. For example, when butter melts in a pan, it is changing from a solid to a liquid. When you fill an ice cube tray with water and put it in the freezer, the water changes from a liquid to a solid. Such changes of state do not change the chemical makeup of matter—just its physical form. As you will have noticed, changes of state are related to temperature. Matter in a liquid state needs to be cooled to change to

As water vapor in the air cools, its molecules move closer together, or condense, to form droplets of liquid water. This is how clouds form.

> ### SUBLIMATION
>
> A few substances, such as carbon dioxide, can change directly from a solid to a gas without becoming a liquid. This is called **sublimation.** Below -109°F (-78.5°C) carbon dioxide is a solid, which is often called dry ice. Above that temperature, dry ice turns into, or sublimes into, carbon dioxide (CO_2) gas. It does not go through a liquid stage.

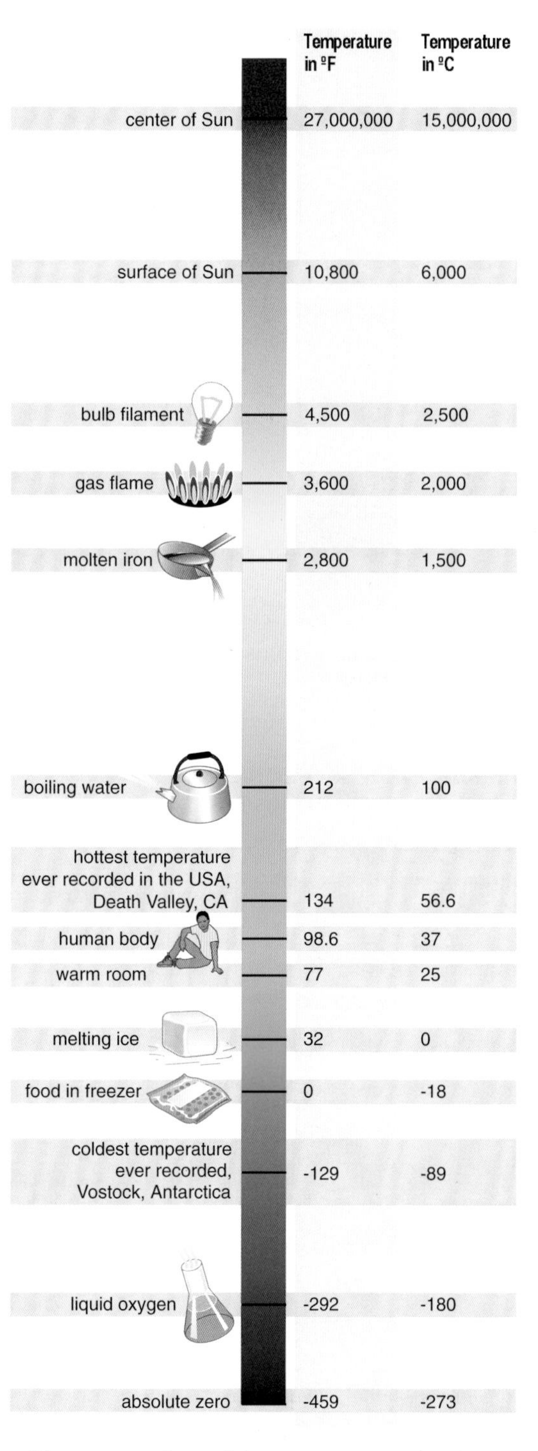

The range of possible temperatures begins at absolute zero (-459°F or -273°C) and rises up to many thousands of degrees.

NAMING THE CHANGES

There are four main ways for matter to change states. They are:

Melting: A solid changes to a liquid
Freezing: A liquid changes to a solid
Evaporation: A liquid changes to a gas
Condensation: A gas changes to a liquid

The **boiling point** is the temperature at which a substance changes from a liquid to a gas. The **condensation point** is the temperature at which a substance changes from a gas to a liquid. The **melting point** is the temperature at which a substance changes from a solid to a liquid. The **freezing point** is the temperature at which a substance changes from a liquid to a solid. Boiling and condensation points are the same, and freezing and melting points are the same. For example water boils and **condenses** at 212°F (100°C) and melts and freezes at 32°F (0°C).

a solid state, and it needs to be heated to change to a gaseous state. The temperatures necessary for such a change depend upon the properties of that particular substance.

The kinetic theory

Have you ever wondered why changes of state happen, or why the three states of matter exist? One of the best explanations we have so far is called the kinetic theory.

Keep in mind that the particles within atoms are constantly moving. The energy present in this type of movement is called **kinetic energy**. According to this theory, as a substance is heated, the energy within the molecules increases. The temperature of a substance is the measure of kinetic energy of its particles: Increasing the temperature increases the movement. The more energy that molecules have, the more they move around and the easier it is for them to break away from each other. So heating a solid turns it first into a liquid (as the vibrating molecules detach from each other) and then into a gas (as the molecules break free).

Cooling down

The kinetic theory works in reverse, too. As a substance cools down, its molecules lose energy and their movement slows down. They move closer together, and the matter changes first from a gas into a liquid and then from a liquid into a solid.

In addition, for some types of matter, there is sublimation, in which a solid changes directly into a gas (see page 22).

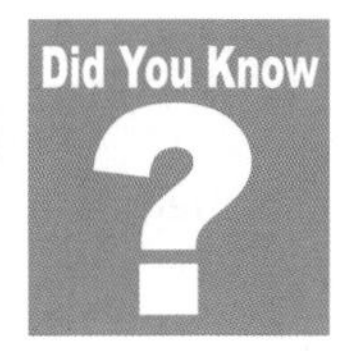

Water is one of the kinds of matter that can turn directly from a solid to a gas, or sublime. If you leave ice cubes in the freezer for a few months, they will shrink, because some of the ice molecules break away to become water vapor (gas).

Density, mass, and volume

If you were given a wooden ball 6 inches (15 centimeters) in diameter, you could carry it without any trouble. But if you had a solid gold ball of the same size, you would hardly be able to lift it. Why are some substances heavier than others? The answer has to do with three more properties of matter: **density**, mass, and **volume**.

How heavy a substance feels depends on its density. Density is a measure of the mass of a substance (how much matter there is in it) compared to its volume (how much space it takes up). For example, in solid gold, matter is very tightly packed. That makes gold a dense matter.

A gold ball 6 inches (15 cm) in diameter would have a mass of 75 pounds (34 kilograms)—about the same density as the average ten-year-old child. A ball made of pine wood would be far less dense—that is, its molecules would not be as tightly packed. It would take up the same amount of space, but it would have a mass of about 1.5 pounds (706 grams).

What is weight?

Weight is a measure of the pull of **gravity** on objects. All matter has gravity, which means it pulls other matter toward it. Earth is a huge object with lots of mass, so it has very strong gravity. Objects with more mass get pulled harder and feel heavier. While mass and density stay the same, weight depends on how strong gravity is. On the Moon, for example, things weigh less than on Earth, because the Moon's gravity is weaker. In outer space, objects don't weigh anything. They are in free fall, either in orbit around a planet, or between other objects in space.

The weights a weightlifter uses are made of dense metals, such as cast-iron and steel.

WATER IS DIFFERENT

Unlike most substances, water gets more dense when it changes from a solid (ice) to a liquid. Ice cubes float in water because they are less dense than water. This happens because water has an unusual property. When it freezes and becomes solid ice, its molecules line up in such a way that they are actually farther apart than they are in liquid water—just the opposite of what happens with most other substances. So, the solid state of water is less dense than the liquid state!

HOW DO YOU CALCULATE DENSITY?

To measure the density of a substance, you first need to measure its mass and volume. That will give you the information necessary to use the formula: density = mass ÷ volume (density equals mass divided by volume). For example, gold has a density of 10.98 ounces per cubic inch (19.3 grams per cubic centimeter).

Changing density

As substances change state between solids, liquids, and gases, their density changes, too. When a solid melts into a liquid, it usually becomes less dense as its particles spread out more. As a liquid becomes a gas, it becomes even less dense.

Measuring mass

Scientists measure the mass of an object by comparing it with a known mass. One way to do this is to use balance scales with weights on one side and the object being measured on the other side. If liquid or gas is being measured, it can be held in a container. Then the mass of the container is deducted from the total.

Measuring volume

To measure the volume of a liquid, you can pour it into a measuring cup, graduated cylinder, or other such container.

To measure the volume of a solid, you can drop it into liquid in a measuring container and measure how much the volume of the liquid increases by being displaced by the solid. The volume of displaced water will be exactly the same as the volume of the solid.

The volume of a gas can also be measured by determining how much water it displaces. This method of measuring volume uses the same basic idea as placing a solid object into water and recording the change in water level. In the case of measuring the gas, however, the gas forces the water out of a second container that originally was full of water.

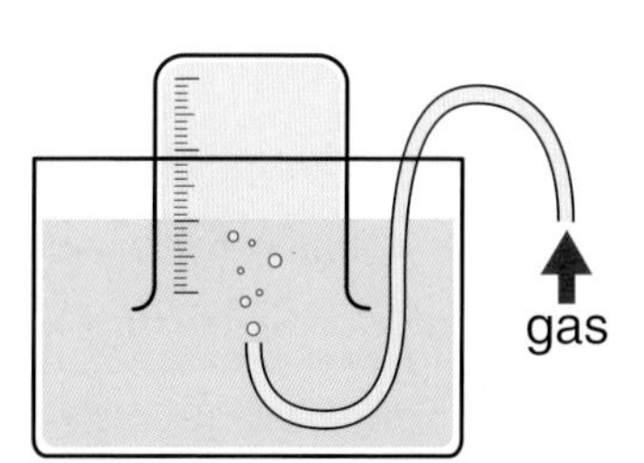

Matter and Energy

Energy is the ability to do work. That could mean the energy you have to run up the stairs or throw a ball. But energy also comes in other forms, such as light, heat, sound, and electricity. Matter is also a form of energy.

Chemical energy

Chemical energy is energy that's stored in matter and released during a chemical reaction. For example, coal contains chemical energy. When you burn coal, the chemical energy changes into heat and light energy. Food also contains chemical energy. After you eat it, chemical reactions happen inside your body and turn the chemical energy into kinetic energy (movement energy) in your muscles.

Potential energy

You can use matter to store and release energy. For example, if you pick up a ball and carry it to the top of a hill, it has

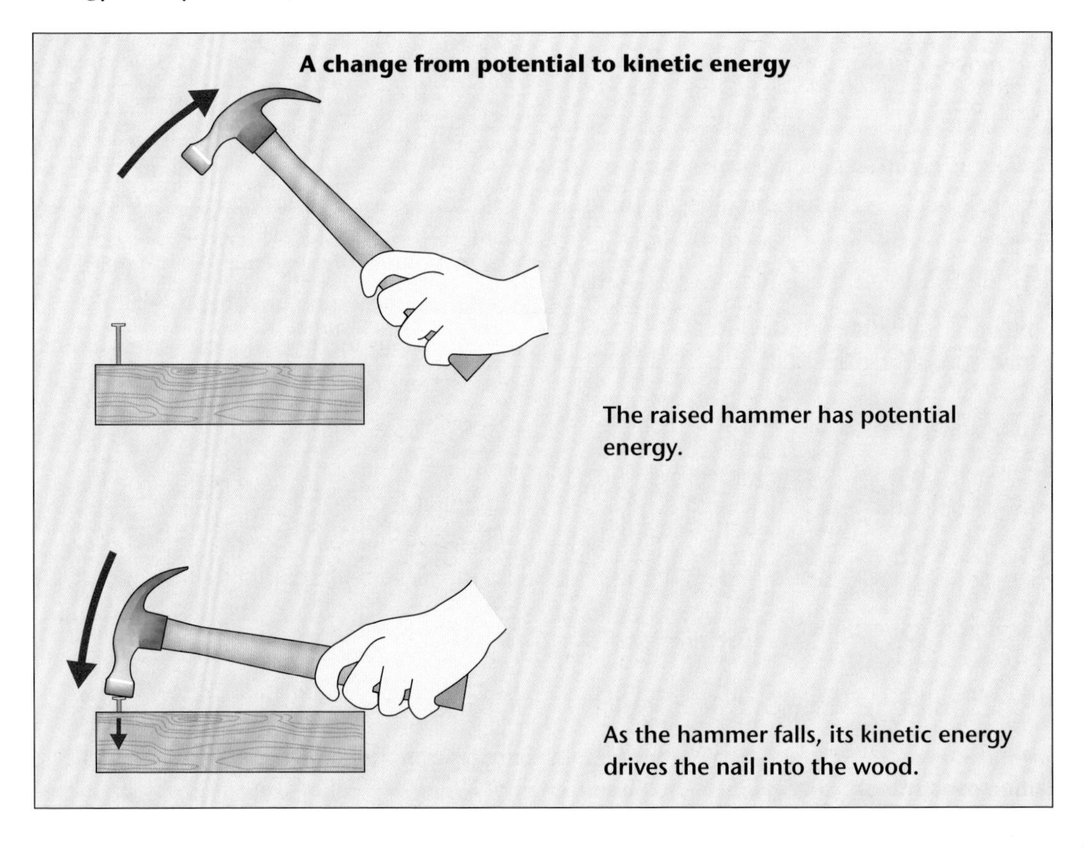

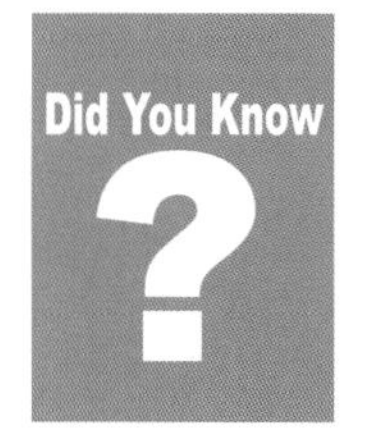

Fat is a very efficient source of energy for the body. One ounce (about 28 grams) of fat provides twice as much energy as the same amount of a sugar or a protein. The human body is also very good at storing the extra fat that we eat. That is why eating a balanced diet is important for staying in good physical shape.

potential energy. As the ball rolls down the hill, its potential energy is converted into kinetic energy.

Energy swaps

All the time, all around us, energy is being stored in matter and then released. This is how we live our lives—using fuels to heat our homes and run cars and trains, eating food so we can live and move, and using our energy to move objects.

CALORIES

We use **calories** as a measurement of the amount of chemical energy in food. For example, 3.5 ounces (100 grams) of cheese has about 400 calories, while 3.5 ounces (100 g) of carrots has about 41 calories.

The more calories a food has, the more energy it gives your body. If you don't use all the energy, it's stored in your body as fat—which is a kind of storage for chemical energy.

Food contains energy, but the amount of energy in a particular food depends upon what the food is made of and how much of it there is. Pizza, for example, is particularly high in carbohydrates and fat, so it contains a lot of calories. The same amount of fresh fruit or vegetables alone contains far fewer calories and far less energy.

In the natural world, matter and energy are constantly affecting each other and converting into each other in cycles and chains of activity. An interchange between matter and energy occurs when rain falls, rivers flow, and plants grow.

Energy cycles

In an energy cycle, a pattern of changes occurs many times. The water cycle is an example. Heat from the Sun changes water in the seas and oceans from a liquid to a gas—water vapor—through the process of evaporation. The water vapor rises into the air, giving it potential energy. As it cools, the water vapor condenses, turning back into liquid water, and falls as rain. The water's potential energy is converted into kinetic energy as the rain falls and streams and rivers flow downhill. Finally, the water

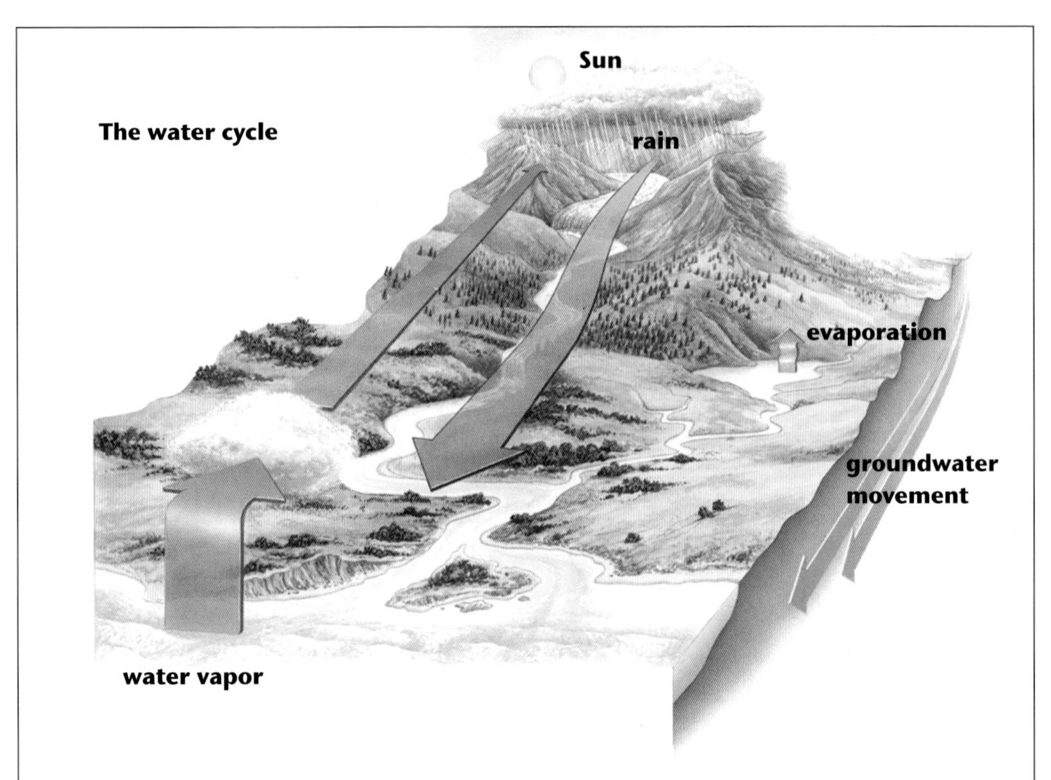

Heat from the Sun gives kinetic energy to water molecules on the surface of oceans, lakes, and even puddles. This causes the water to evaporate and change into water vapor. Water vapor rises, gaining potential energy. Water vapor cools and condenses, becoming liquid water. Liquid water falls as rain, releasing kinetic energy. The rain runs into streams and rivers or soaks into the ground. Eventually it returns to the sea as part of a flowing river, stream, or groundwater, and the cycle continues.

ALBERT EINSTEIN

German physicist Albert Einstein (1879–1955) was one of the greatest scientists who ever lived. He was rebellious and did badly at school but went on to make many important discoveries about the nature of matter, light, time, and space. In 1905, Einstein came to the realization that matter is a form of energy. Matter and energy, he said, are interchangeable. His famous equation, $E = mc^2$, explains the relationship between matter and energy. This equation means that the amount of energy (E) in a given amount of mass (m) is equal to the mass times the speed of light (c), squared.

flows back into the seas and oceans, and the cycle begins again.

Energy chains

In an energy chain, energy moves along in a sequence of changes. A food chain is an example of an energy chain. Plants use light from the Sun to make their own food so they can build their stems, leaves, roots, flowers, fruits, seeds, and other parts. They convert light energy into matter for food, using a process called **photosynthesis**. Animals eat the plants, turning chemical energy from the plants into kinetic energy. Animals also use the energy from plants to build their own bodies. Plant-eating animals (herbivores), in turn, are eaten by meat-eating animals, or carnivores. Each time this happens, some of the chemical energy stored in the body of the herbivore is absorbed by the carnivore. This again turns chemical energy into kinetic energy. Animals also give off energy in the form of heat and sound.

The food chain

This simple food chain has four links. The flowers are eaten by beetles, which are eaten by rats, which are eaten by owls. In this way, energy is passed through the food chain from the flowers to each of the animals.

EINSTEIN ON MATTER

"Time and space and gravitation have no separate existence from matter."

Albert Einstein

Moving through matter

Matter allows some forms of energy to pass through it. When light shines through your window, light energy is passing through the glass. You hear sounds because sound energy is carried through the air. And heat energy spreads out through matter—for example, when sunshine warms a lake.

Sound vibrations

Sound energy can travel through the air and many other substances, too. This happens because sound travels by vibrating particles of matter in particular patterns. As the particles vibrate, they bump into other particles, and the patterns of vibrations spread out. When they hit our ears, we hear sounds. Sound can only travel through matter, such as air or water.

Ice and water are transparent.

Light

Some substances are transparent, meaning that they let nearly all the light that hits them pass through. Some are translucent, allowing some, but not all, light to pass through.

Other substances are opaque, which means they don't let light through at all. The molecules of an opaque substance absorb (soak up) light.

When light strikes a molecule in a transparent substance, such as a pane of clear glass, the energy of the glass molecule increases slightly, so the glass might feel warm. The light energy itself keeps traveling, however, until the light exits the other side of the window. This process slows the light a little, making it bend, or **refract**.

Heat

According to the kinetic theory (see page 23), heat makes particles in matter move around more. A hot object will pass its heat

THE SPEED OF SOUND

In air, sound travels at about 761 miles per hour (1,225 kilometers per hour)—that's quite fast. In a denser substance, such as water, it travels faster, because the particles are closer together and transfer the sound vibrations more quickly. The speed of sound in water is 3,375 miles per hour (5,400 kph)—that's more than four times faster than in air!

WHAT IS REFRACTION?

Light slows down as it travels through a transparent substance, such as glass or water. If a beam of light enters and leaves the substance at an angle, one edge of the beam is slowed down first, making the light bend. Bending is called refraction, and it can change the way things appear. If you stand in the water and look down, your feet look nearer than they are. This occurs because light reflected from your feet bends, or refracts, as it leaves the water. If you put a coin in a glass full of water, you should be able to look at the glass from just above one side without seeing the coin. The water refracts, or bends, the light, which makes the coin seem to disappear.

onto a cooler object until both objects are the same temperature. That's why a hot potato for example, gradually cools down as it passes its heat to the surrounding air. If you touch a hot iron, your hand burns, as the iron passes its heat to you.

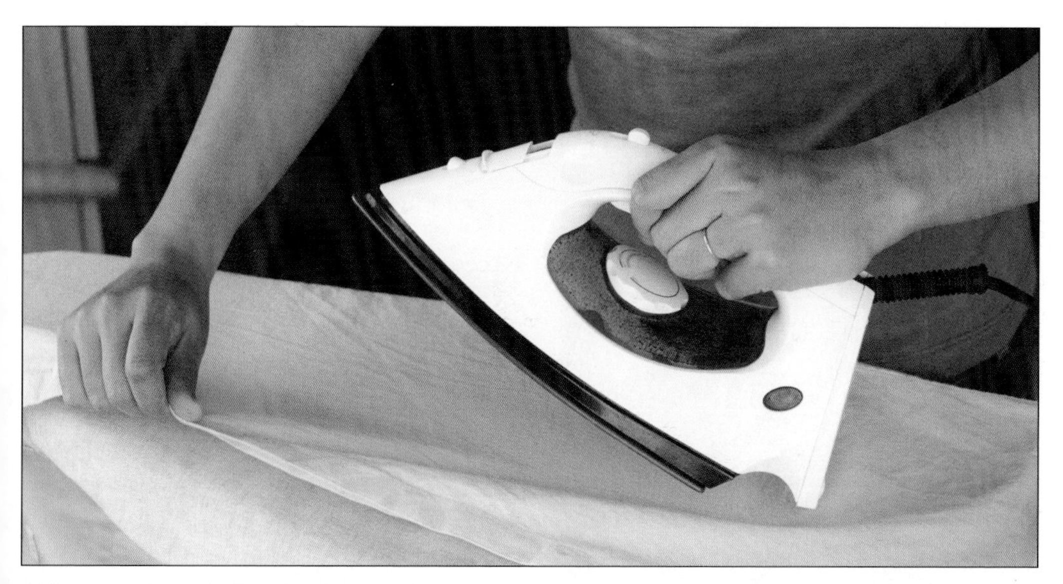

When you iron clothes, heat passes from the iron into the fabric. The heat changes the fabric's shape and smoothes out any wrinkles.

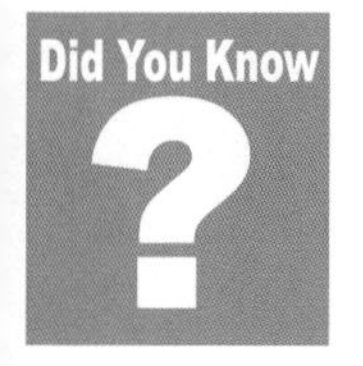

Like sound, light and heat can travel through matter. But unlike sound, they also can travel through a vacuum—a completely empty space. Heat and light from the Sun travel across millions of miles of empty space to reach us on Earth.

Properties of Substances

Matter can be separated into two categories: substances and mixtures. A substance is a particular type of matter that is composed of certain ingredients combined in a certain way. A mixture is a combination of substances that can be separated physically. Unlike substances, the ingredients in a mixture keep their own properties.

Mixtures

A substance may be a single element, such as gold or iron, or it may be a combination of elements, such as water or salt or steel. Each substance has its own properties—those qualities and features that help distinguish one substance from another. We have already discussed some properties such as state of matter, density, mass, and volume, but there are more.

Brass, a mixture of copper and zinc, is used to make brass musical instruments, tools, and furniture. A mixture of metals is called an **alloy**.

The soil in your backyard is a mixture: It may have silt, small rocks, sand, and organic material such as bits of dead leaves and grass in it. A soft drink is a mixture of water, flavorings, sweeteners, and carbon dioxide (the fizz). A mixture can be a combination of elements, substances, or both.

Sometimes it's easy to tell when something is a mixture, and sometimes it's not. If the parts of a mixture are noticeably different from each other, or if they aren't evenly mixed, like a bowl of vegetable soup or the soil in your yard, it's called a **heterogeneous** mixture. If the parts are so similar and so well mixed together that it's hard to tell them apart, that's a **homogeneous** mixture—like your soft drink.

Millions of materials

Between mixtures and substances, today's world puts millions of different materials at our fingertips. The list begins with natural materials such as cork, wool, cotton, and different types of metal, wood, and

THE MOHS SCALE OF HARDNESS

German mineralogist Friedrich Mohs (1773–1839) invented this scale to measure the hardness of materials.

Hardness number		Example mineral
1	talc	
2	gypsum	
3	calcite	
4	fluorite	
5	apatite	
6	orthoclase	
7	quartz	
8	topaz	
9	corundum	
10	diamond	

stone. In addition, scientists have developed synthetic materials—substances and mixtures—that are produced in laboratories or factories. These include plastics, nonstick coatings, synthetic fabrics such as nylon, and stainless steel (which is a homogeneous mixture of different metals).

Some substances, such as gold, are composed of single elements. Others substances consist of molecules made up of two or more elements. Polyethylene, for example, contains molecules made of hydrogen and carbon atoms. Still others are mixtures in which, as discussed earlier, the elements or compounds retain their individual properties. For example, brass is a mixture of two metals, zinc and copper. Such a mixture of metals—called an alloy—combines some of the most desirable properties of each of the metals, such as hardness, durability, and beauty.

WHICH IS THE HARDEST?

The diamond is the hardest natural material known. A diamond is so hard that it can only be cut by another diamond.

A list of properties

The list of properties of any material are qualities that are measurable and observable. Here are just a few:

- Hardness: the ability to resist damage, wear, and tear

- Flexibility: how easily a material bends
- Elasticity: how much a material can be stretched or squashed, yet still return to its original shape
- Conductivity: how well a material can conduct (carry) electricity and heat when compared to other materials
- Strength: the ability to withstand stress.

Strength can be measured in many ways:

tensile strength—strength when stretched
compressive strength—strength when squashed
torsional strength—strength when twisted
shear strength—strength when cut
bending strength—strength when bent
malleability—how easily a material can be beaten and reshaped

Testing materials

You may want to collect a sample of materials from around your home or school and examine them. That's one way to understand just how many different properties they have. Substances you could examine might include wooden objects, cork, a metal coin, a paperclip, a piece of chalk, a plastic pen lid, sponge, a pebble, a piece of cheese, cotton fabric, tissue paper, cardboard, an eraser, a cherry stone, a piece of candle—and anything else you can find. Remember: Everything is made up of matter.

Tests you could try include:

- Does it float in water?
- Can it be squashed?
- If it can be squashed, does it spring back into shape?
- Does it stick to a magnet?
- Does it bend or snap?
- Can it be cut with a knife?
- Does it soak up water?
- What happens if you put it in the freezer for a day?
- Does light shine through it?

How hard is it?

- Soft: can be scratched with a fingernail
- Medium: can be scratched with a knife or glass, but not with a fingernail
- Hard: cannot be scratched by a knife

You could give each material a score from one to ten in each test and enter all the results into your own testing chart.

Metalworkers and shipbuilders have long known that certain forms of copper can discourage the growth of living matter, such as germs, algae (seaweed), and tiny shellfish (barnacles). Because of that property, brass, an alloy of copper and zinc, is commonly used for doorknobs and other hardware, and copper-based paints and metal facings are used on the hulls of boats and ships.

TEST FOR HEAT CONDUCTION

This experiment tests how well a material conducts heat. You need a wooden spoon, a metal spoon, and a plastic spoon, butter, frozen peas, hot water, and three cups. Each of the cups, spoons, and peas should be as similar in size and shape as possible to the others of its group.

1. Fill the cups with hot, almost-boiling water.

2. Put a small lump of butter on the end of each spoon handle. Stick a frozen pea into the butter on each spoon.

3. Now stand each spoon in a cup of hot water, with the bowl of the spoon in the water and the handle sticking out. Put them in all at the same time.

4. Time how long it takes for the peas to fall off the spoons. The better a material conducts heat, the faster heat will travel up the spoon, and the sooner the pea will fall off.

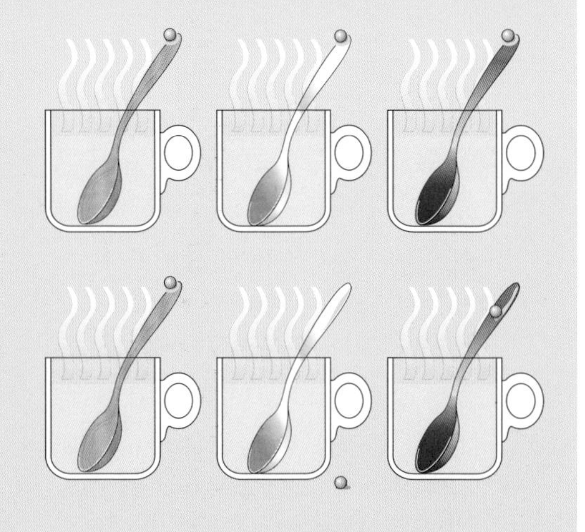

Professional testing

Believe it or not, many people have full-time jobs testing materials, though they use specially designed equipment to do their testing. Materials have to be tested to make sure they are safe and suitable for whatever they are going to be used for. For example, concrete for building a bridge will undergo lots of tests on its strength and flexibility.

In this photo, materials used to make soccer balls are being tested by machines that compress a test ball.

Using Materials

We use different materials for different jobs, depending on their properties. Looking at the properties of various materials makes it easy to understand why some materials are better suited for some jobs than for others.

Fit for the job

All the objects we use have to be made of the right materials, so that they work well, last well, and are safe to use. For example, a steel saucepan is strong, and conducts heat well. It has a high melting point, so it stays solid even over a fierce flame. It is waterproof, durable (lasts a long time), and not too heavy to lift. All these things are important for it to do its job.

Some materials have similar properties, so they can be used for the same purpose. For example, glass and clear plastic both let light through, so they could both be used to make eyeglasses, transparent packaging, cell phone display windows, wristwatch covers, and oven doors. Can you think of why manufacturers might choose one material over another for different purposes?

From where do materials come?

To make the many products we use every day, we have to collect materials from the world around us. For example, materials such as gold, rubies, and emeralds are used for making jewelry; granite and marble are used for statues and buildings; gypsum is used in plaster, toothpaste, and cement. All these materials are found under the ground. We get them by mining. Wood, rubber, and cork come from trees; cotton comes from a plant; and wool comes from sheep. We get these materials by farming and ranching.

Materials taken from nature are known as raw materials. They usually have to be processed (treated and changed) before they can be used in many of the products we know.

Cork, used for making bottle stoppers, floor tiles, and many other products, is a natural material that comes from the bark of a cork oak tree.

Stone Age

The earliest people lived in the Stone Age (250,000–10,000 B.C.), which has now been divided into more specific periods known as Paleolithic and Neolithic. Humans living in the lower Paleolithic period (250,000,000–200,000 B.C.) made tools by chipping and carving stone. The mineral flint, for example, was used to shape ax heads, blades, and other useful items.

Bronze Age

The Bronze Age began between 4000 and 3000 B.C. At that time, people in Asia, Africa, and Europe began using metals from the ground to make tools. They mainly used copper or bronze. Bronze, which is copper mixed with tin, gave the age its name.

Iron Age

The Iron Age began somewhere around 1000 B.C. People in this time period learned to

Everyday objects are often made of many different materials. For example, cell phones, seen here being made in a factory, include plastic, rubber, and several different types of metal.

MATERIAL AGES

People didn't always have access to all the materials we have today. Long ago, humans used whatever materials they could find. Human history is divided into time periods, called ages, based on the materials people used.

The precious metal platinum, used in jewelry and in some machines, is one of Earth's rarest raw materials. We recycle as much platinum as we can, because it is so expensive. Scientists estimate that all the platinum ever mined would fit inside a cube measuring about 20 feet (6 meters) on each side. How does that cube compare to the size of your classroom?

extract iron, a much stronger metal than copper or tin, from rocks found in the ground.

New materials

Many of the materials we use today aren't found in nature. We invented them by processing and combining other substances. For example, paper is made by chopping and mashing wood into a pulp, mixing it with water and other chemicals, and rolling it into a flat sheet. Plastics, which we use to make thousands of everyday things, are synthetic materials made by processing chemicals found in crude oil.

Inventing materials

Scientists are constantly developing new materials to make life easier, do jobs better, or allow new objects to be invented. Every so often, they come up with a new material that has lots of uses and becomes part of modern life. One well-known example is Kevlar. American inventor Stephanie Kwolek (b.1923), working for the DuPont Company, invented this extra-tough material in 1966. It is a **polymer**—a material made up of chain-shaped molecules. Kevlar is very strong, heat-resistant, and difficult to cut, pierce, or break.

The Mars Exploration Rover *Opportunity*, seen here being tested on Earth, used Kevlar airbags to protect it as it descended onto the surface of Mars in 2004.

KEVLAR—A STRONG MATERIAL WITH MANY USES

Here are just a few of the many uses of super-strong Kevlar:

- bulletproof vests
- strong, lightweight skis
- sports helmets
- heat-proof gloves
- tennis rackets
- reinforcement in airplanes
- puncture-proof tire linings
- extra-strong cables, including those that are the only thing attaching the basket (and passengers, fuel, and burners) to the balloon part of a hot air balloon.
- kayaks
- the tether cable used to lower the Mars *Pathfinder* lander onto the surface of Mars in 1997
- bomb disposal suits
- motorcycle gear

An army bomb disposal expert wears a protective Kevlar suit during bomb disposal training.

STEPHANIE KWOLEK

Stephanie Kwolek (b.1923) is the U.S. scientist who invented Kevlar. She began work at the materials company DuPont in 1946 and worked on using chemical reactions to create new types of polymers (substances made of chain-shaped molecules). When she first developed Kevlar, it had no special qualities, but she found that when it was spun into thin strands, it had enormous strength. Besides Kevlar, Kwolek patented sixteen other discoveries during her career. She retired in 1986.

Matter Science

Scientists are still trying to find out more about the true nature of matter. Although they have done a lot of research to date and continue to do more, there is still much to know about matter and how it works.

Inside atoms

Although scientists have found out that atoms are made up of smaller particles, they still aren't sure how they all work. Perhaps more subatomic particles are yet to be discovered. One of the main things physicists do is to continue studying atoms to try to determine what they are made of.

Matter experiments

To study atoms, scientists use a machine called a particle accelerator. It works by firing a subatomic particle, such as an electron, at an atom. This breaks it apart so that scientists can find out what parts it is made up of. To do this, the particle has to travel very fast—almost as fast as the speed of light, which is 186,000 miles (300,000 km) per second. Some particle accelerators fire particles down a long tube; others shoot them into a huge ring where they pick up speed as they move around the ring. The biggest particle accelerator built to date is in a laboratory, CERN (from the French name **C**onseil **E**uropéen pour la **R**echerche **N**ucléaire,

In this aerial photo, the red line shows the path of the underground, ring-shaped CERN particle accelerator, near Geneva, Switzerland. This circle is 5 miles (8.5 km) across.

STRANGE MATTER

Some scientists study the way different types of matter behave. Their tests help them determine if they can control matter—and make different types do unusual things. In 2003 for example, scientists at the University of Texas in Austin found that if they vibrated a liquid mixture of water and cornstarch at a certain speed, they could poke holes in it, just as if it were a solid.

or European Council for Nuclear Research), near Geneva, Switzerland.

When atoms join together instead of splitting, it's called **nuclear fusion**. Nuclear fusion happens inside the Sun, as atoms collide and join together to make new atoms. As they do so, they give out heat. Now, scientists are trying to develop another kind of nuclear power based on nuclear fusion. If they can do it, it should be safer and cheaper than **nuclear fission** power.

NUCLEAR CHANGES

"The release of atom power has changed everything. . . If only I had known, I should have become a watchmaker."

Albert Einstein, regretting that his discoveries contributed to the development of nuclear bombs.

NUCLEAR FISSION

In the 1930s, scientists found a way to split atoms of the element uranium apart. As each atom split, it released a great deal of heat energy. It also sent out particles that split open nearby atoms. They in turn split other atoms, causing a chain reaction.

Splitting the nucleus of atoms is called nuclear fission. It is the basis of nuclear power and nuclear weapons. If the splitting happens in a slow, controlled way, the heat can be collected and used as an energy source. If it is uncontrolled, it causes a devastating explosion—a nuclear bomb.

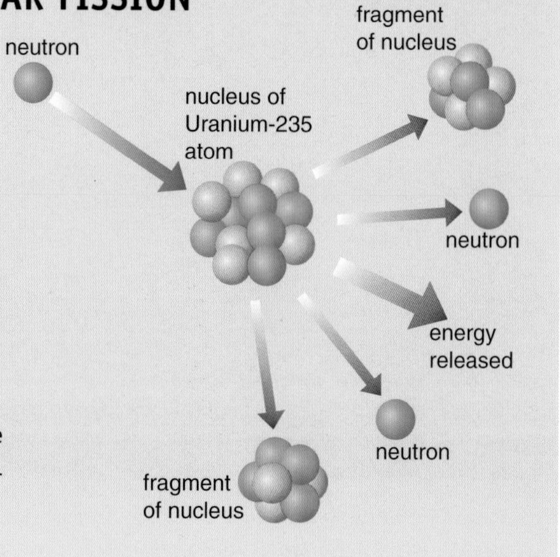

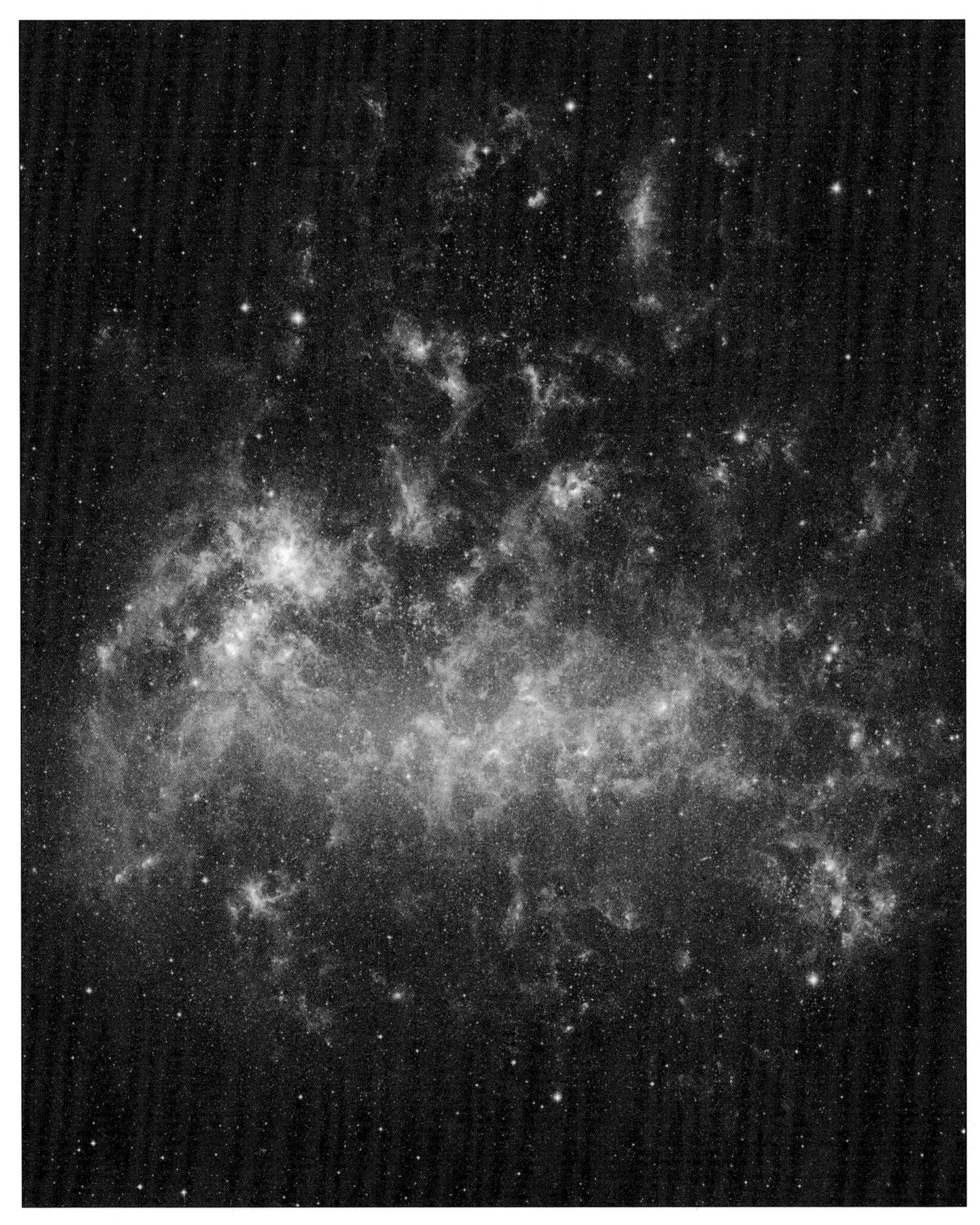

All the stars, planets, and comets, and indeed everything in the universe, such as this galaxy, is made of matter. Scientists are investigating exactly how and why the universe began.

Matter mysteries

Matter is at the heart of the biggest questions scientists are trying to answer. By understanding matter, we can understand more about what ourselves, our surroundings, and the whole universe are really made of. In the meantime, scientists continue to tackle questions, including those in the following list.

Puzzling gravity

We know that gravity exists because matter pulls other matter toward it. But how? If you jump up in the air, Earth's gravity pulls you down, even though Earth isn't touching you. It's still a mystery how matter can have a pulling force across empty space.

String theory

Most scientists say that the smallest particles of matter are like tiny dots. But a few think that particles behave like tiny strings. This is called string theory, and scientists continue to debate about it.

NANOTECHNOLOGY

Nanotechnology is a new area of science that involves making microscopic machines, structures, and tools measuring just a few atoms across. These tools have many possible uses, such as carrying medicines into the body, keeping food free of bugs, and making self-cleaning clothes.

Big Bang

Another big question asks where matter came from in the first place. Many scientists think the whole universe started with a massive explosion called the "big bang." Before that, they think, there was no space, time, or matter at all. But we still don't know why the big bang happened, or what ultimately will happen to the universe.

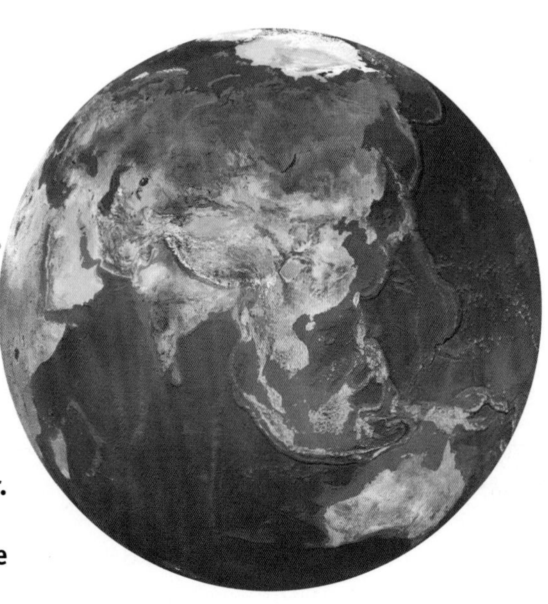

Planet Earth was formed from the dust and rock of old, exploded stars. In turn, all living things on Earth are made from the planet's matter. So you, your friends, and all humans are made of stardust!

GLOSSARY

alloy a mixture of metals

atom one of the tiny units from which everything is made. The atom is the smallest single part of an element, containing tiny particles around a central nucleus.

atomic mass unit (amu) the amount of mass in one neutron or one proton

atomic number the number of protons an atom has in its nucleus

boiling point temperature at which a substance changes from a liquid to a gas

bond a union of two or more atoms to make a molecule

calorie a unit of energy, often used to measure the amount of energy in food processed by the body

chemical energy energy stored in a substance such as a food or fuel

chemical reaction what happens when substances swap and rearrange their atoms to become new substances

chemistry the science of how atoms bond together and make substances

compound a substance made up of the atoms of two or more elements bonded together to make molecules

condensation the change from a gas into a liquid

condensation point the temperature at which a substance changes from a gas to a liquid

density the amount of matter in a substance compared to its volume

electric charge a kind of pulling force in an atom or particle, which can be positive or negative

electron a negatively charged type of subatomic particle that flies around the nucleus of an atom

element a substance made up of just one type of atom

elemental molecule a molecule made of atoms of the same substance bonded together

energy the ability to do work

evaporation the change from a liquid into a gas. This change takes place at the liquid's surface.

freezing changing from a liquid into a solid

freezing point temperature at which a substance changes from a liquid to a solid

gas a state of matter in which molecules are widely spread out and a substance has no fixed shape

gravity a force that causes matter to be pulled toward other matter

group a column of elements in the Periodic Table

heterogeneous a description for a mixture that is composed of identifiably different substances and parts

homogeneous a description for a mixture that has the same structure and composition throughout

ion a charged atom

isotopes atoms of the same element that have different numbers of neutrons

kinetic energy energy that takes the form of movement

liquid a state of matter in which molecules are loosely attached; a substance that can pour, flow, and take the shape of its container

mass the amount of matter in an atom, object, or substance

mass number the total number of protons and neutrons an atom has in its nucleus

melt to change from a solid into a liquid

melting point temperature at which a substance changes from a solid to a liquid

mixture a substance in which two or more elements or compounds come together but retain their individual properties

molecule the smallest whole unit of a substance that retains all of the properties of the original

molecular model a model that shows how atoms are joined together to make a molecule

neutron a neutral subatomic particle found in the nucleus of an atom

nuclear fission splitting the nucleus of an atom, releasing energy

nuclear fusion joining two atomic nuclei together, releasing energy

nucleus (plural **nuclei**) the center of an atom, containing protons and neutrons

particle a tiny part

period a row of elements in the Periodic Table

Periodic Table of the Elements also known as the Periodic Table, is a scientific charting of all known elements

photosynthesis a process in plants that uses energy from sunlight to make food

physics the study of matter and how it works

polymer a type of substance made up of chain-shaped molecules

potential energy energy stored in an object because of its position, such as a ball at the top of a slope

properties qualities or features of matter that can be observed and measured

proton a positively charged subatomic particle found in the nucleus of an atom

refract to bend because of a change in speed when moving from one substance to another

relative atomic mass (RAM) the average mass number of the atoms in an element

solid a state of matter in which molecules are firmly held together, giving substances and objects a fixed shape

states of matter the three main forms in which matter can exist: solid, liquid, and gas

subatomic particles particles that make up atoms

sublimation changing directly from a solid into a gas

synthetic artificial or made by humans

theory a possible explanation of how something works, based on careful observations, experiments, or study

universe a name for everything that exists, including all of space

vibrate to shake back and forth in a regular pattern

volume the amount of space that an object or substance takes up

weight a measure of how strongly gravity pulls on an object

Books

Arnold, Nick. *Chemical Chaos* (Horrible Science series). New York: Scholastic, 1998.

Brady, James E., and Senese, Fred. *Chemistry: Matter and Its changes*. Indianapolis: Wiley, 2004.

Cooper, Christopher. *Matter* (Eyewitness series). New York: Dorling Kindersley, 1999.

Shevick, Edward. *Physical Science: Matter and motion* (Science Action Labs series). Carthage, IL: Teaching & Learning Company, 1998.

Solway, Andrew. *Tall Tales and Mad Scientists: Atoms and Elements* (Raintree Fusion series). Chicago: Raintree, 2005.

VanCleave, Janice. *Janice VanCleave's Chemistry for Every Kid: 101 Experiments that Really Work.* Indianapolis: Jossey-Bass, 1989.

Walker, Sally M., and Andy King. *Matter* (Early Bird Energy series). Minneapolis: Lerner Publications, 2005.

Web sites

Chem4Kids.com

www.chem4kids.com/files/matter_intro.html
Useful guide to all kinds of chemistry and matter facts.

8th Grade Sci-ber text: Matter

www.usoe.k12.ut.us/curr/science/sciber00/8th/matter/sciber/intro.htm Information, experiments, videos, and other resources to do with matter science.

Adventures in Science and Technology

collections.ic.gc.ca/science/english/index.html Experiments, facts, and profiles of real working scientists. Just click on *physics* or *chemistry* to go to the right sections.

Schoolscience

www.schoolscience.co.uk/content/index.asp Information summaries, experiments, diagrams and quizzes to help with science subjects.

Web sites (cont.)

Strange Mattter

www.strangematterexhibit.com/

Interactive games, experiments, close-up views of matter, and useful facts.

The Particle Adventure

particleadventure.org/particleadventure/

An interactive tour of atoms, particles, and matter mysteries.

Fermilabyrinth

ed.fnal.gov/projects/labyrinth/games/index.html

Fun page of games and discoveries from the Fermilab particle physics laboratory.

Viscosity explorer

www.seed.slb.com/en/scictr/lab/visco_exp/index.htm

Test the viscosity (resistance to movement) of different liquids with this online experiment.

Publisher's note to educators and parents: Our editors have carefully reviewed these Web sites to ensure that they are suitable for children. Many Web sites change frequently, however, and we cannot guarantee that a site's future contents will continue to meet our high standards of quality and educational value. Be advised that children should be closely supervised whenever they access the Internet.

Places to visit

American Museum of Science and Energy
300 S. Tulane Avenue
Oak Ridge, TN 37830
865-576-3200
www.amse.org/

Miami Museum of Science and Planetarium
3280 South Miami Avenue
Miami, FL 33129
305-646-4200
www.miamisci.org

The Exploratorium
(San Francisco's world-famous science museum)
3601 Lyon Street
San Francisco, CA 94123
415-EXP-LORE
www.exploratorium.edu/

Arizona Science Center
600 East Washington Street
Phoenix, Arizona 85004
602-716-2000
www.azscience.org/

INDEX